OUR PURPOSE

THE
NOBEL PEACE PRIZE
LECTURE

2007

Al Gore

RODALE

Portions of this book were previously published by Rodale Inc. as
An Inconvenient Truth © 2006.

© 2008 by Al Gore

Rodale books may be purchased for business or promotional use or for
special sales. For information, please write to:

Special Markets Department, Rodale, Inc., 733 Third Avenue,
New York, NY 10017

Printed in Canada

Rodale Inc. makes every effort to use acid-free ∞, recycled paper ♻.

Book design by Darlene Schneck

Front cover photo © Don Farrall/Getty Images

Library of Congress Cataloging-in-Publication Data

Gore, Albert, date
    Our purpose : the Nobel Peace Prize lecture 2007 / Al Gore.
        p.   cm.
    ISBN-13 978–1–60529–990–7 paperback
    ISBN-10 1–60529–990–1 paperback
    1. Gore, Albert, date   2. Environmentalists—United States.
3. Nobel Prize winners—United States.   4. Speeches, addresses, etc.
5. Nobel Prizes.   6. Peace—Awards.   I. Title.
GE56.G67 2008
363.7—dc22   2008003921

Distributed to the book trade by Macmillan

2 4 6 8 10 9 7 5 3 1   paperback

We inspire and enable people to improve their lives and the world around them

For more of our products visit **rodalestore.com** or call 800-848-4735

OUR PURPOSE

THE
NOBEL PEACE PRIZE
LECTURE

OSLO, 2007

Speech by Al Gore
on the Acceptance of the
Nobel Peace Prize

December 10, 2007
Oslo, Norway

Your Majesties, Your Royal Highnesses, Honorable members of the Norwegian Nobel Committee, Excellencies, ladies and gentlemen.

I have a purpose here today. It is a purpose I have tried to serve for many years. I have prayed that God would show me a way to accomplish it.

Sometimes, without warning, the future knocks on our door with a precious and painful vision of what might be. One hundred and nineteen years ago, a wealthy inventor read his own obituary, mistakenly published years before his death. Wrongly believing the inventor had just died, a newspaper printed a harsh judgment of his life's work, unfairly labeling him "The Merchant of Death" because of his invention—dynamite. Shaken by this condemnation, the inventor made a fateful choice to serve the cause of peace.

Seven years later, Alfred Nobel created this prize and the others that bear his name.

Seven years ago tomorrow, I read my own political obituary in a judgment that seemed to me harsh and mistaken—if not premature. But that unwelcome verdict also brought a precious if painful gift: an opportunity to search for fresh new ways to serve my purpose.

Unexpectedly, that quest has brought me here. Even though I fear my words cannot match this moment, I pray what I am feeling in my heart will be communicated clearly enough that those who hear me will say, "We must act."

The distinguished scientists with whom it is the greatest honor of my life to share this award have laid before us a choice between two different futures—a choice that to my ears echoes the words of an ancient prophet: "Life or death, blessings or curses. Therefore, choose life, that both thou and thy seed may live."

We, the human species, are confronting a planetary emergency—a threat to the survival of our civilization that is gathering ominous and destructive potential even as we gather here. But there is hopeful news as well: We have the ability to solve this crisis and avoid the worst—though not all—of its consequences, if we act boldly, decisively and quickly.

However, despite a growing number of honorable exceptions, too many of the world's leaders are still best described in the words Winston Churchill applied to those who ignored Adolf Hitler's threat: "They go on in strange paradox, decided only to be undecided, resolved to be irresolute, adamant for drift, solid for fluidity, all powerful to be impotent."

So today, we dumped another 70 million tons of global-warming pollution into the thin shell of atmosphere surrounding our planet, as if it were an open sewer. And tomorrow, we will dump a slightly larger amount, with the cumulative concentrations now trapping more and more heat from the sun.

As a result, the Earth has a fever. And the fever is rising. The experts have told us it is not a passing affliction that will heal by itself. We asked for a second opinion. And a third. And a fourth. And the consistent conclusion, restated with unprecedented alarm, is that something basic is wrong.

We are what is wrong, and we must make it right.

Last September 21, as the Northern Hemisphere tilted away from the sun, scientists reported with increasing distress that the

North Polar ice cap is "falling off a cliff." One study estimated that it could be completely gone during summer in less than 22 years. Another new study, to be presented by U.S. Navy researchers later this week, warns it could happen in as little as 7 years.

Seven years from now.

In the last few months, it has been harder and harder to misinterpret the signs that our world is spinning out of kilter. Major cities in North and South America, Asia, and Australia are nearly out of water due to massive droughts and melting glaciers. Desperate farmers are losing their livelihoods. Peoples in the frozen Arctic and on low-lying Pacific Islands are planning evacuations of places they have long called home. Unprecedented wildfires have forced a half million people from their homes in one country and caused a national emergency that almost brought down the government in another. Climate refugees have migrated into areas already inhabited by people with different cultures, religions, and traditions, increasing the potential for conflict. Stronger storms in the Pacific and Atlantic have threatened whole cities. Millions have been displaced by massive flooding in South Asia, Mexico, and 18 countries in Africa. As temperature extremes have increased, tens of thousands have lost their lives. We are recklessly burning and clearing our forests and driving more and more species into extinction. The very web of life on which we depend is being ripped and frayed.

We never intended to cause all this destruction, just as Alfred Nobel never intended that dynamite be used for waging war. He had hoped his invention would promote human progress. We shared that same worthy goal when we began burning massive quantities of coal, then oil and methane.

Even in Nobel's time, there were a few warnings of the likely consequences. One of the very first winners of the prize in chemistry worried that, "We are evaporating our coal mines into the air." After performing 10,000 equations by hand, Svante Arrhenius calculated that the Earth's average temperature would increase by many degrees if we doubled the amount of CO_2 in the atmosphere.

Seventy years later, my teacher, Roger Revelle, and his colleague, Dave Keeling, began to precisely document the increasing CO_2 levels day by day.

But unlike most other forms of pollution, CO_2 is invisible, tasteless, and odorless—which has helped keep the truth about what it is doing to our climate out of sight and out of mind. Moreover, the catastrophe now threatening us is unprecedented—and we often confuse the unprecedented with the improbable.

We also find it hard to imagine making the massive changes that are now necessary to solve the crisis. And when large truths are genuinely inconvenient, whole societies can, at least for a time, ignore them. Yet as George Orwell reminds us: "Sooner or later a false belief bumps up against solid reality, usually on a battlefield."

In the years since this prize was first awarded, the entire relationship between humankind and the Earth has been radically transformed. And still, we have remained largely oblivious to the impact of our cumulative actions.

Indeed, without realizing it, we have begun to wage war on the Earth itself. Now, we and the Earth's climate are locked in a relationship familiar to war planners: "Mutually assured destruction."

More than 2 decades ago, scientists calculated that nuclear war could throw so much debris and smoke into the air that it would block life-giving sunlight from our atmosphere, causing a "nuclear winter." Their eloquent warnings here in Oslo helped galvanize the world's resolve to halt the nuclear arms race.

Now science is warning us that if we do not quickly reduce the global warming pollution that is trapping so much of the heat our planet normally radiates back out of the atmosphere, we are in danger of creating a permanent "carbon summer."

As the American poet Robert Frost wrote, "Some say the world will end in fire; some say in ice." Either, he notes, "would suffice."

But neither need be our fate. It is time to make peace with the planet.

We must quickly mobilize our civilization with the urgency and resolve that has previously been seen only when nations mobilized for war. These prior struggles for survival were won when leaders found words at the 11th hour that released a mighty surge of courage, hope, and readiness to sacrifice for a protracted and mortal challenge.

These were not comforting and misleading assurances that the threat was not real or imminent; that it would affect others but not ourselves; that ordinary life might be lived even in the presence of extraordinary threat; that Providence could be trusted to do for us what we would not do for ourselves.

No, these were calls to come to the defense of the common future. They were calls upon the courage, generosity, and strength of entire peoples, citizens of every class and condition who were ready to stand against the threat once asked to do so. Our enemies in those times calculated that free people would not rise to the challenge; they were, of course, catastrophically wrong.

Now comes the threat of climate crisis— a threat that is real, rising, imminent, and universal. Once again, it is the 11th hour. The penalties for ignoring this challenge are immense and growing and, at some near point, would be unsustainable and unrecoverable. For now we still have the power to choose our fate, and the remaining question is only this: Have we the will to act vigorously and in time, or will we remain imprisoned by a dangerous illusion?

Mahatma Gandhi awakened the largest democracy on Earth and forged a shared resolve with what he called *satyagraha*, or "truth force."

In every land, the truth—once known—has the power to set us free.

Truth also has the power to unite us and bridge the distance between "me" and "we," creating the basis for common effort and shared responsibility.

There is an African proverb that says, "If you want to go quickly, go alone. If you want to go far, go together." We need to go far, quickly.

We must abandon the conceit that individual, isolated, private actions are the answer. They can and do help. But they will not take us far enough without collective action. At the same time, we must ensure that in mobilizing globally, we do not invite the establishment of ideological conformity and a new lock-step "ism."

That means adopting principles, values, laws, and treaties that release creativity and initiative at every level of society in multifold responses originating concurrently and spontaneously.

This new consciousness requires expanding the possibilities inherent in all humanity. The innovators who will devise a new way to harness the sun's energy for pennies or invent an engine that's carbon negative may live in Lagos or Mumbai or Montevideo. We must ensure that entrepreneurs and inventors everywhere on the globe have the chance to change the world.

When we unite for a moral purpose that is manifestly good and true, the spiritual energy unleashed can transform us. The generation that defeated fascism throughout the world in the 1940s found, in rising to meet their awesome challenge, that they had gained the moral authority and long-term vision to launch the Marshall Plan, the United Nations, and a new level of global cooperation and foresight that unified Europe and facilitated the emergence of democracy and prosperity in Germany, Japan, Italy, and much of the world. One of their visionary leaders said, "It is time we steered by the stars and not by the lights of every passing ship."

In the last year of that war, you gave the Peace Prize to a man from my hometown of 2,000 people, Carthage, Tennessee. Cordell Hull was described by Franklin Roosevelt as the "Father of the United Nations." He was an inspiration and hero to my own father, who followed Hull in the Congress and the U.S. Senate and in his commitment to world peace and global cooperation.

My parents spoke often of Hull, always in tones of reverence and admiration. Eight weeks ago, when you announced this prize, the deepest emotion I felt was when I saw the headline in my hometown paper that simply noted I had won the same prize that Cordell Hull had won. In that moment, I knew what my father and mother would have felt were they alive.

Just as Hull's generation found moral authority in rising to solve the world crisis caused by fascism, so too can we find our greatest opportunity in rising to solve the climate crisis. In the kanji characters used in both Chinese and Japanese, "crisis" is written with two symbols—the first meaning "danger," the second, "opportunity."

By facing and removing the danger of the climate crisis, we have the opportunity to gain the moral authority and vision to vastly increase our own capacity to solve other crises that have been too long ignored.

We must understand the connections between the climate crisis and the afflictions of poverty, hunger, and HIV/AIDS and other pandemics. As these problems are linked, so too must be their solutions. We must begin by making the common rescue of the global environment the central organizing principle of the world community.

Fifteen years ago, I made that case at the Earth Summit in Rio de Janeiro. Ten years ago, I presented it in Kyoto. This week, I will urge the delegates in Bali to adopt a bold mandate for a treaty that establishes a universal global cap on emissions and uses the market in emissions trading to efficiently allocate resources to the most effective opportunities for speedy reductions.

This treaty should be ratified and brought into effect everywhere in the world by the beginning of 2010—two years sooner than presently contemplated. The pace of our response must be accelerated to match the accelerating pace of the crisis itself.

Heads of state should meet early next year to review what was accomplished in Bali and take personal responsibility for addressing this crisis. It is not unreasonable to ask, given the gravity of our circumstances, that these heads of state meet every three months until the treaty is completed.

We also need a moratorium on the construction of any new generating facility that burns coal without the capacity to safely trap and store carbon dioxide.

And most important of all, we need to put a price on carbon—with a CO_2 tax that is then rebated back to the people, progressively, according to the laws of each nation, in ways that shift the burden of taxation from employment to pollution. This is by far the most effective and simplest way to accelerate solutions to this crisis.

The world needs an alliance—especially of those nations that weigh heaviest in the scales where Earth is in the balance. I salute Europe and Japan for the steps they've taken in recent years to meet the challenge, and the new government in Australia, which has made solving the climate crisis its first priority.

But the outcome will be decisively influenced by two nations that are now failing to do enough: the United States and China. While India is also growing fast in importance, it should be absolutely clear that it is the two largest CO_2 emitters—most of all, my own country—that will need to make the boldest moves or stand accountable before history for their failure to act.

Both countries should stop using the other's behavior as an excuse for stalemate and instead develop an agenda for mutual survival in a shared global environment.

These are the last few years of decision, but they can be the first years of a bright and hopeful future if we do what we must. No one should believe a solution will be found without effort, without cost, without change. Let us acknowledge that if we wish to redeem squandered time and speak again with moral authority, then these are the hard truths:

The way ahead is difficult. The outer boundary of what we currently believe is feasible is still far short of what we actually must do. Moreover, between here and there, across the unknown, falls the shadow.

That is just another way of saying that we have to expand the boundaries of what is possible. In the words of the Spanish poet, Antonio Machado, "Pathwalker, there is no path. You must make the path as you walk."

We are standing at the most fateful fork in that path. So I want to end as I began, with a vision of two futures—each a palpable possibility—and with a prayer that we will see with vivid clarity the necessity of choosing between those two futures, and the urgency of making the right choice now.

The great Norwegian playwright, Henrik Ibsen, wrote, "One of these days, the younger generation will come knocking at my door."

The future is knocking at our door right now. Make no mistake, the next generation will ask us one of two questions. Either they will ask: "What were you thinking; why didn't you act?"

Or they will ask instead: "How did you find the moral courage to rise and successfully resolve a crisis that so many said was impossible to solve?"

We have everything we need to get started, save perhaps political will, but political will is a renewable resource.

So let us renew it and say together: "We have a purpose. We are many. For this purpose, we will rise and we will act."

Excerpts from
An Inconvenient Truth

IN EVERY CORNER OF THE GLOBE—on land and in water, in melting ice and disappearing snow, during heat waves and droughts, in the eyes of hurricanes and in the tears of refugees—the world is witnessing mounting and undeniable evidence that nature's cycles are profoundly changing.

I have learned that, beyond death and taxes, there is at least one absolutely indisputable fact: Not only does human-caused global warming exist, but it is also growing more and more dangerous, and at a pace that has now made it a planetary emergency.

Since my childhood summers on our family's farm in Tennessee, when I first learned from my father about taking care of the land, I have been deeply interested in learning more about threats to the environment. I grew up half in the city and half in the country, and the half I loved most was on our farm. Since my mother read to my sister and me from Rachel Carson's classic book, *Silent Spring*, and especially since I was first introduced to the idea of global warming by my college professor Roger Revelle, I have always tried to deepen my own understanding of the human impact on nature, and in my public service I have tried to implement policies to ameliorate—and eventually eliminate—that harmful impact.

After the 2000 election, one of the things I decided to do was to start giving my slide show on global warming again. I had first put it together at the same time I began writing

Earth in the Balance, and over the years I have added to it and steadily improved it to the point where I think it makes a compelling case that humans are the cause of most of the global warming that is taking place, and that unless we take quick action the consequences for our planetary home could become irreversible.

For the last six years, I have been traveling around the world, sharing the information I have compiled with anyone who would listen in colleges, small towns, and big cities. More and more, I have begun to feel that I am changing minds, but it is a slow process.

Although it is true that politics at times must play a crucial role in solving this problem, this is the kind of challenge that ought to completely transcend partisanship. So whether you are a Democrat or a Republican, whether you voted for me or not, I very much hope that you will sense that my goal is to share with you both my passion for the Earth and my deep sense of concern for its fate. It is impossible to feel one without the other when you know all the facts.

I also want to convey my strong feeling that what we are facing is not just a cause for alarm, it is paradoxically also a cause for hope. As many know, the Chinese expression for "crisis" consists of two characters side by side. The first is the symbol for "danger," the second, the symbol for "opportunity."

The climate crisis is, indeed, extremely dangerous. In fact it is a true planetary emergency. Two thousand scientists, in a hundred countries, working for more than 20 years in the most elaborate and well-organized scientific collaboration in the history of humankind, have forged an exceptionally strong consensus that all the nations on Earth must work together to solve the crisis of global warming.

The voluminous evidence now strongly suggests that unless we act boldly and quickly to deal with the underlying causes of global warming, our world will undergo a string of terrible catastrophes, including more and stronger storms like Hurricane Katrina, in both the Atlantic and the Pacific.

We are melting the North Polar ice cap and virtually all of the mountain glaciers in the world. We are destabilizing the massive mound of ice on Greenland and the equally enormous mass of ice propped up on top of islands in West Antarctica, threatening a worldwide increase in sea levels of as much as 20 feet.

The list of what is now endangered due to global warming also includes the continued stable configuration of ocean and wind currents that has been in place since before the first cities were built almost 10,000 years ago.

We are dumping so much carbon dioxide into the Earth's environment that we have literally changed the relationship between the Earth and the Sun. So much of that CO_2 is being absorbed into the oceans that if we continue at the current rate we will increase the saturation of calcium carbonate to levels that will prevent formation of corals and interfere with the making of shells by any sea creature.

Global warming, along with the cutting and burning of forests and other critical habitats, is causing the loss of living species at a level comparable to the extinction event that wiped out the dinosaurs 65 million years ago. That event was believed to have been caused by a giant asteroid. This time it is not an asteroid colliding with the Earth and wreaking havoc; it is us.

Last year, the national academies of science in the 11 most influential nations came together to jointly call on every nation to "acknowledge that the threat of climate change is clear and increasing" and declare that the "scientific understanding of climate changes is now sufficiently clear to justify nations taking prompt action."

So the message is unmistakably clear. This crisis means "danger!"

Why do our leaders seem not to hear such a clear warning? Is it simply that it is inconvenient for them to hear the truth?

If the truth is unwelcome, it may seem easier just to ignore it.

But we know from bitter experience that the consequences of doing so can be dire.

For example, when we were first warned that the levees were about to break in New Orleans because of Hurricane Katrina, those warnings were ignored. Later, a bipartisan group of members of Congress chaired by Representative Tom Davis (R-VA), chairman of the House Government Reform Committee, said in an official report, "The White House failed to act on the massive amounts of information at its disposal," and that a "blinding lack of situational awareness and disjointed decision-making needlessly compounded and prolonged Katrina's horror."

Today, we are hearing and seeing dire warnings of the worst potential catastrophe in the history of human civilization: a global climate crisis that is deepening and rapidly becoming more dangerous than anything we have ever faced.

And yet these clear warnings are also being met with a "blinding lack of situational awareness"—in this case, by the Congress, as well as the president.

As Martin Luther King Jr. said in a speech not long before his assassination:

"We are now faced with the fact, my friends, that tomorrow is today. We are confronted with the fierce urgency of now. In this unfolding conundrum of life and history, there is such a thing as being too late.

"Procrastination is still the thief of time. Life often leaves us standing bare, naked, and dejected with a lost opportunity. The tide in the affairs of men does not remain at flood—it ebbs. We may cry out desperately for time to pause in her passage, but time is adamant to every plea and rushes on. Over the bleached bones and jumbled residues of numerous civilizations are written the pathetic words 'Too late.' There is an invisible book of life that faithfully records our vigilance or our neglect. Omar Khayyam is right: 'The moving finger writes, and having writ moves on.'"

But along with the danger we face from global warming, this crisis also brings unprecedented opportunities.

What are the opportunities such a crisis also offers? They include not just new jobs and new profits, though there will be plenty of both, we can build clean engines, we can harness the Sun and the wind; we can stop wasting energy; we can use our planet's plentiful coal resources without heating the planet.

The procrastinators and deniers would have us believe this will be expensive. But in recent years, dozens of companies have cut emissions of heat-trapping gases while saving money. Some of the world's largest companies are moving aggressively to capture the enormous economic opportunities offered by a clean energy future.

But there's something even more precious to be gained if we do the right thing.

The climate crisis also offers us the chance to experience what very few generations in history have had the privilege of knowing: a generational mission; the exhilaration of a compelling moral purpose; a shared and unifying cause; the thrill of being forced by circumstances to put aside the pettiness and conflict that so often stifle the restless human need for transcendence; the opportunity to rise.

When we do rise, it will fill our spirits and bind us together. Those who are now suffocating in cynicism and despair will be able to breathe freely. Those who are now suffering from a loss of meaning in their lives will find hope.

When we rise, we will experience an epiphany as we discover that this crisis is not really about politics at all. It is a moral and spiritual challenge.

At stake is the survival of our civilization and the habitability of the Earth. Or, as one eminent scientist put it, the pending question is whether the combination of an opposable thumb and a neocortex is a viable combination on this planet.

The understanding we will gain—about who we really are—will give us the moral capacity to take on other related challenges that are also desperately in need of being redefined as moral imperatives with practical solutions: HIV/AIDS and other pandemics that are ravaging so many; global poverty; the ongoing redistribution of wealth globally from the poor to the wealthy; the ongoing genocide in Darfur; the ongoing famine in Niger and elsewhere; chronic civil wars; the destruction of ocean fisheries; families that don't function; communities that don't commune; the erosion of democracy in America; and the refeudalization of the public forum.

Consider what happened during the crisis of global fascism. At first, even the truth about Hitler was inconvenient. Many in the west hoped the danger would simply go away. They ignored clear warnings and compromised with evil, and waited, hoping for the best.

After the appeasement at Munich, Churchill said: "This is only the first sip, the first foretaste of a bitter cup which will be proffered to us year by year—unless by supreme recovery of moral health and martial vigor, we rise again and take our stand for freedom."

But when England and then America and our allies ultimately rose to meet the threat, together we won two wars simultaneously in Europe and the Pacific.

By the end of that terrible war, we had gained the moral authority and vision to create the Marshall Plan—and convinced taxpayers to pay for it! We had gained the spiritual capacity and wisdom to rebuild Japan and Europe and launch the renewal of the very nations we had just defeated in war, in the process laying the foundation for 50 years of peace and prosperity.

This too is a moral moment, a crossroads. This is not ultimately about any scientific discussion or political dialogue. It is about who we are as human beings. It is about our capacity to transcend our own limitations, to rise to this new occasion. To see with our hearts, as well as our heads, the response that is now called for. This is a moral, ethical, and spiritual challenge.

We should not fear this challenge. We should welcome it. We must not wait.

In the words of Dr. King, "Tomorrow is today."

•

Many people today still assume—mistakenly—that the Earth is so big that we human beings cannot possibly have any major impact on the way our planet's ecological system operates. That assertion may have been true at one time, but it's not the case anymore. We have grown so numerous and our technologies have become so powerful that we are now capable of having a significant influence on many parts of the Earth's environment. The most vulnerable part of the Earth's ecological system is the atmosphere. It's vulnerable because it's so thin.

My friend, the late Carl Sagan, used to say, "If you had a globe covered with a coat of varnish, the thickness of that varnish would be about the same as the thickness of the Earth's atmosphere compared to the Earth itself."

The atmosphere is thin enough that we are capable of changing its composition.

Indeed, the Earth's atmosphere is so thin that we have the capacity to dramatically alter the concentration of some of its basic molecular components. In particular, we have vastly increased the amount of carbon dioxide—the most important of the so-called greenhouse gases.

Under normal conditions, a portion of outgoing infrared radiation is naturally trapped by the atmosphere—and that is a good thing, because it keeps the temperature on Earth within comfortable bounds. The greenhouse gases on Venus are so thick that its temperatures are far too hot for humans. The greenhouse gases surrounding Mars are almost nonexistent, so the temperature there is far too cold. That's why the Earth is sometimes referred to as the "Goldilocks planet"—the temperatures here have been just right.

The problem we now face is that this thin layer of atmosphere is being thickened by huge quantities of human-caused carbon dioxide and other greenhouse gases. And as it thickens, it traps a lot of the infrared radiation that would otherwise escape the atmosphere and continue out to the universe. As a result, the temperature of the Earth's

atmosphere—and oceans—is getting danger-
ously warmer.

That's what the climate crisis is all about.

•

Many people say about the rising temperatures, "Oh, it's just natural variability.
These things go up and down, so we shouldn't
worry."

As the oceans get warmer, storms get
stronger. In 2004, Florida was hit by four
unusually powerful hurricanes.

A growing number of new scientific studies
are confirming that warmer water in the top
layer of the ocean can drive more convection
energy to fuel more powerful hurricanes.

As water temperatures go up, wind veloc-
ity goes up, and so does storm moisture
condensation.

The science textbooks had to be rewritten
in 2004. They used to say, "It's impossible to

have hurricanes in the South Atlantic." But that year, for the first time ever, a hurricane hit Brazil.

Hard on the heels of 2004 came the record-breaking summer of 2005. Several hurricanes hit the Caribbean and the Gulf of Mexico early in the season, including Hurricane Dennis and Hurricane Emily, which caused significant damage.

And then came Katrina. When it first hit Florida on its way into the Gulf on the morning of August 26, 2005, it was only a category 1 storm, but it killed a dozen people and caused billions of dollars in damage.

Then, it passed over the unusually warm waters of the Gulf of Mexico. By the time Katrina hit New Orleans, it was a massive and powerfully destructive storm.

The consequences were horrendous. There are no words to describe them.

●

There are two places on Earth that serve as canaries in the coal mine—regions that are especially sensitive to the effects of global warming. The first is the Arctic. The second is the Antarctic. In both of these frozen realms, scientists are seeing faster changes and earlier, more dramatic effects of climate change than anywhere else on Earth.

In photographs, these two ends of the Earth superficially resemble one another. In both places, ice and snow are everywhere you look. But beneath the surface, there is a dramatic difference between them. In contrast to the massive, 10,000-foot-thick Antarctic ice cap, the Arctic ice cap is, on average, less than 10 feet thick.

And beneath the ice at each pole lies the reason for the difference: The Antarctic is land surrounded by ocean, while the Arctic is ocean surrounded by land.

The sheer thinness of the Arctic's floating ice—and of the frozen layer of soil in the land area north of the Arctic Circle surrounding the Arctic Sea—makes it highly vulnerable to the sharply rising temperatures.

As a result, the most dramatic impact of global warming in the Arctic is the accelerated melting. Temperatures are shooting upward there faster than at any other place on the planet.

Since the 1970s, the extent and thickness of the Arctic ice cap has diminished precipitously. There are now studies showing that if we continue with business as usual, the Arctic ice cap will completely disappear each year during the summertime. At present, it plays a crucial role in cooling the Earth. Preventing its disappearance must be one of our highest priorities.

What does it mean to us to look at a vast expanse of open water, at the top of our world, that used to be—but is no longer—covered by ice? We ought to care about this a lot, because it has serious planetary effects.

The average temperature worldwide is about 58°F.

An increase of five degrees actually means an increase of only one or two degrees at the Equator, but more than 12° at the North Pole, and a large increase on the periphery of Antarctica as well.

And so all those wind and ocean current patterns that formed during the last ice age, which have been relatively stable ever since, are now up in the air.

Our civilization has never experienced any environmental shift remotely similar to this. Today's climate pattern has existed throughout the entire history of human civilization.

Every place—every city, every farm—is located or has been developed on the basis of the same climate patterns we have always known.

The age-old rhythm of the Earth's seasons—summer, fall, winter, and spring—is also changing as some parts of the world heat up more rapidly than others.

Many species around the world are now threatened by climate change, and some are becoming extinct—in part because of the climate crisis and in part because of human encroachment into the places where they once thrived.

In fact, we are facing what biologists are beginning to describe as a mass extinction crisis, with a rate of extinction now 1,000 times higher than the normal background rate.

•

The second canary in the coal mine—along with the Arctic—is Antarctica, the largest mass of ice on the planet by far.

Antarctica is the closest thing to another planet we can experience on this one.

It is surreal—completely and unremittingly white in every direction, so vast and so cold—much colder than the Arctic. The enormity of all that snow masks a surprising fact: Antarctica is actually a desert. It meets the technical definition in that it receives less than one inch of precipitation per year. Think about it—an icy desert, a freeze-dried oxymoron.

Scientists thought the Larsen-B ice shelf would be stable for at least another century—even with global warming. But starting on January 31, 2002, within 35 days, it completely broke up. Indeed, most of it disappeared over the course of two of those days.

Once the sea-based ice shelf was gone, the land-based ice behind it that was being held back began to shift and fall into the sea. This, too, was unexpected and carries important implications because ice—whether in the form of a mountain glacier or a land-based ice shelf in Antarctica or Greenland—raises the sea level when it melts or falls into the sea.

This is one of the reasons sea levels have been rising worldwide, and will continue to go up if global warming is not quickly checked.

Many residents of low-lying Pacific Island nations have already had to evacuate their homes because of rising seas.

If Greenland melted or broke up and slipped into the sea—or if half of Greenland and half of Antarctica melted or broke up and slipped into the sea, sea levels worldwide would increase by between 18 and 20 feet.

Tony Blair's advisor, David King, is among the scientists who have been warning about the potential consequences of large changes in these ice shelves.

At a 2004 conference in Berlin, he said, "The maps of the world will have to be redrawn."

•

The fundamental relationship between our civilization and the ecological system of the

Earth has been utterly and radically transformed due to the powerful convergence of three factors.

The first is the population explosion, which in many ways is a success story in that death rates and birth rates are going down everywhere in the world, and families, on average, are getting smaller. But even though these hoped-for developments have been taking place more rapidly than anyone would have anticipated a few decades ago, the momentum in world population has built up so powerfully that the "explosion" is still taking place and continues to transform our relationship to the planet.

If you look at population growth in the context of history, it is obvious that the last 200 years represent a complete break with the pattern that prevailed for most of the millennia that humans have walked on the Earth. From the time when scientists say our species first appeared 160,000 to 190,000 years ago, until the time of Jesus Christ and Julius Caesar, human population had grown to a quarter of a billion people. By the time of America's birth in 1776, it had grown to 1 billion. When the baby boom generation that I'm a part of was born at the end of World War II, the population had just crossed 2 billion. In my lifetime, I have watched it go all the way to 6.5 billion. My generation will see it rise to more than 9 billion people.

The point is simple and powerful: It took more than 10,000 generations for the human population to reach 2 billion. Then it began to rocket upward from 2 billion to 9 billion in the course of a single lifetime: ours.

This rapid population rise drives demand for food, water, and energy—and for all our natural resources. It puts enormous pressure on vulnerable areas like forests—particularly the rain forests of the tropics.

And that brings me to the second factor that has transformed our relationship to the Earth—the scientific and technological revolution.

New advances in science and technology have brought us tremendous improvements in areas like medicine and communications, among many others. For all the advantages we have gained from our new technologies, we have also witnessed many unanticipated side effects.

The new power we have at our disposal hasn't always been accompanied by new wisdom in the way we use it, particularly when we exercise our technologically enhanced power in the thoughtless pursuit of age-old habits, which are, after all, hard to change.

The third and final factor causing the collision between humankind and nature is both the subtlest and most important: our fundamental way of thinking about the climate crisis.

And the first problem in the way we think about the climate crisis is that it seems easier not to think about it at all. One reason it doesn't consistently demand our attention can be illustrated by the classic story about an old science experiment involving a frog that jumps into a pot of boiling water and immediately jumps out again because it instantly recognizes the danger. The same frog, finding itself in a pot of lukewarm water that is being slowly brought to a boil, will simply stay in the water—in spite of the danger—until it is . . . rescued.

(I used to recount this story about the frog with a different ending to the last sentence above: "until the frog is boiled." But after dozens of slide shows were followed by at least one anguished listener coming up to me and expressing concern for the fate of the frog, I finally learned the importance of rescuing the frog.)

But of course the larger point of the story is that our collective "nervous system," through which we recognize an impending danger to our survival, is similar to the frog's. If we experience a significant change in our circumstances gradually and slowly, we are capable of sitting still and failing to recognize the seriousness of what is happening to us until it's too late. Sometimes, like the frog, we only react to a sudden jolt, a dramatic and speedy change in our circumstances that sets off our alarm bells.

Global warming may seem gradual in the context of a single lifetime, but in the context of the Earth's history, it is actually happening with lightning speed. Its pace is now accelerating so rapidly that even in our own lifetimes, we are beginning to see the telltale bubbles of a boiling pot.

We are, of course, also different from the frog. We don't have to wait for the boiling point in order to understand the danger we're in—and we do have the ability to rescue ourselves.

The second problem in the way we think about the climate crisis is the wide gulf between what C. P. Snow described as "the two cultures." Science has become so specialized in its single-minded pursuit of ever-more refined knowledge in narrowing subspecialties that the rest of us have more and more difficulty making sense of their conclusions and translating their wisdom into plain language. Moreover, since science thrives on uncertainty and politics is paralyzed by it, scientists have a difficult time sounding the alarm bells for politicians, because even when their findings make it clear that we're in grave danger, their first impulse is to replicate the experiment to see if they get the same results.

Politicians, on the other hand, often confuse self-interested arguments paid for by lobbyists and planted in the popular press with legitimate, peer-reviewed studies published in reputable scientific journals. For example, the so-called global warming skeptics cite one article more than any other in arguing that global warming is just a myth: a statement of concern during the 1970s that the world might be in danger of entering a new ice age. But the article in which that scientist's comment appeared was published in *Newsweek* and never appeared in any peer-reviewed journal. Moreover, the scientist who made the statement corrected it shortly thereafter with a clear explanation of why his offhand comment was erroneous.

There is a misconception that the scientific community is in a state of disagreement about whether global warming is real, whether human beings are the principal cause,

and whether its consequences are so dangerous as to warrant immediate action. In fact,
there is virtually no serious disagreement
remaining on any of these central points that
make up the consensus view of the world
scientific community.

According to Jim Baker, when he was
head of NOAA, the scientific agency respon-
sible for most of the measurements related to
global warming, "There is a better scientific
consensus on this issue than any other . . .
with the possible exception of Newton's Law
of Dynamics."

The misconception that there is serious
disagreement among scientists about global
warming is actually an illusion that has been
deliberately fostered by a relatively small but
extremely well-funded cadre of special inter-
ests, including Exxon Mobil and a few other
oil, coal, and utilities companies. These com-

panies want to prevent any new policies that would interfere with their current business plans that rely on the massive unrestrained dumping of global warming pollution into the Earth's atmosphere every hour of every day.

One of the internal memos prepared by this group to guide the employees they hired to run their disinformation campaign was discovered by the Pulitzer Prize–winning reporter Ross Gelbspan. Here was the group's stated objective: to "reposition global warming as theory, rather than fact."

This technique has been used before.

The tobacco industry, 40 years ago, reacted to the historic Surgeon General's report linking cigarette smoking to lung cancer and other lung diseases by organizing a similar disinformation campaign. One of their memos, prepared in the 1960s, was recently uncovered during one of the lawsuits against the tobacco companies on behalf of the millions of people who have been killed by their product.

It is interesting to read it 40 years later in the context of the ongoing global warming disinformation campaign:

"Doubt is our product, since it is the best means of competing with the 'body of fact' that exists in the mind of the general public. It is also the means of establishing a controversy."—Brown and Williamson Tobacco Company memo, 1960s

The third problem in our way of thinking about global warming is our false belief that we have to choose between a healthy economy and a healthy environment.

One of the keys to solving the climate crisis involves finding ways to use the powerful force of market capitalism as an ally. And more than anything else, that requires accurate measurements of the

real consequences—positive and negative—
of all the important economic choices
we make.

Unfortunately, the false choice posed
between our economy and the environment
affects our policies in harmful ways.

Our outdated environmental standards
are based on faulty thinking about the true
relationship between the economy and the
environment. They are intended in this case
to help automobile companies succeed.
But it's the companies building more efficient
cars that are doing well. The U.S. companies
are in deep trouble. And they're still redou-
bling their efforts to sell large, inefficient gas-
guzzlers even though the marketplace is
sending the same message that the environment is sending.

The fourth and final problem in the way
some people think about global warming is the

dangerous misconception that if it really is as big a threat as the scientists are telling us it is, then maybe we're helpless to do anything about it so we might as well throw up our hands.

An astonishing number of people go straight from denial to despair, without pausing on the intermediate step of saying, "We can do something about this!"

And we can.

Each one of us is a cause of global warming, but each of us can become part of the solution: in the decisions we make on what we buy, the amount of electricity we use, the cars we drive, and how we live our lives. We can even make choices to bring our individual carbon emissions to zero.

Ultimately, the question comes down to this: Are we, as Americans, capable of doing great things, even though they might be difficult?

Are we capable of transcending our own

limitations and rising to take responsibility for charting our own destiny?

Well, the record indicates that we have this capacity.

We fought a revolution and brought forth a new nation, based on liberty and individual dignity.

We made a moral decision that slavery was wrong, and that we could not be half-free and half-slave.

We recognized that women should have the right to vote.

We won two wars against fascism simultaneously, in the Atlantic and the Pacific, and then we won the peace that followed.

We cured fearsome diseases like polio and smallpox, improved the quality of life, and extended our lifetimes.

We took on the moral challenge of desegregation and passed civil rights laws to remedy injustice against minorities.

We landed on the Moon—one of the most inspiring examples of what we can do when we put our minds to it.

We have even solved a global environmental crisis before. The problem of the hole in the stratospheric ozone layer was said to be impossible to fix because the causes were global and the solution required cooperation from every nation in the world. But the United States took the lead—on a bipartisan basis—with a Republican president and a Democratic Congress.

We drafted a treaty, secured world-wide agreement on it, and began to eliminate the chemicals that were causing the problem.

Now it is up to us to use our democracy and our God-given ability to reason with one another about our future and make moral choices to change the policies and behaviors that would, if continued, leave a degraded, diminished, and hostile planet for our children and grandchildren—and for humankind.

We must choose instead to make the 21st century a time of renewal. By seizing the opportunity that is bound up in this crisis, we can unleash the creativity, innovation, and inspiration that are just as much a part of our human birthright as our vulnerability to greed and pettiness. The choice is ours. The responsibility is ours. The future is ours.

One of the robotic spacecrafts that America launched years ago to explore the universe took a picture as it was leaving Earth's gravity, a picture of our planet spinning slowly in the void. Years later, when the same spacecraft had traveled 4 billion miles beyond our solar system, the late Carl Sagan suggested that NASA send a signal instructing the craft to turn its cameras toward the Earth again, and from that unimaginable distance, take another photograph of Earth. Sagan called it a pale blue dot and noted that everything that has ever happened in all of human history has happened on that tiny pixel. All the triumphs and tragedies. All the wars. All the famines. All the major advances.

It is our only home.

And that is what is at stake. Our ability to live on planet Earth—to have a future as a civilization.

I believe this is a moral issue.

It is our time to rise again to secure our future.